YOUR KNOWLEDGE HAS VALUE

- We will publish your bachelor's and
  master's thesis, essays and papers

- Your own eBook and book -
  sold worldwide in all relevant shops

- Earn money with each sale

Upload your text at www.GRIN.com
and publish for free

Bibliographic information published by the German National Library:

The German National Library lists this publication in the National Bibliography; detailed bibliographic data are available on the Internet at http://dnb.dnb.de .

Imprint:

Copyright © 2016 GRIN Verlag, Open Publishing GmbH
Print and binding: Books on Demand GmbH, Norderstedt Germany
ISBN: 9783668352315

This book at GRIN:

http://www.grin.com/en/e-book/345132/intelligent-analysis-of-hurricane-data-over-gis-applications

Yuniel Proenza

Intelligent Analysis of Hurricane Data over GIS Applications

GRIN Publishing

Intelligent Analysis of Hurricane Data over GIS Applications

Abstract — Geographical Information Systems (GIS) research area have been evolving with time. Those systems have become useful beyond spatial and geographic information representation and computer aided analysis using maps. Some of the most important fields of application for GIS are fleet control, tourism analysis and Meteorological analysis. For the last mentioned field, advisory and prediction models should be enhanced aiming the avoidance of critical damage associated to Hurricanes. Intelligent systems have become the optimal solution when decision-making situations are extreme and advanced reasoning is expected. Existing models involve high mathematical analysis, which is complex for humans and also sometimes computational costs have to be considered. There are some artificial intelligence fields that are exploring the possibility of making inference about existing data, enriching the information and enhancing the obtained results. In the present work we propose a hybrid GIS which includes some behavioral aspects of intelligent systems for Hurricane analysis.

Keywords – GIS, Hurricane Analysis, Intelligent Systems, Ontology, Movement Semantics

CONTENT

I. INTRODUCTION

The analysis of hurricanes and their impact in determinate areas is mandatory if we have to preserve human life and economic resources. To precisely predict how will evolve and behave one tropical storm in time is the key for meteorologists on this venue. Even we have precise mathematic and geo-graphic models for understanding this phenomena, the diverse nature and the high number of variables that interact in the life cycle of one hurricane makes extremely difficult to say we are exact in prediction, even more with a minimum probability of failure. The answer lies in the "understanding of hurricanes as they mean", taking into account what origin them, which factors have influence in their evolution and behavior and what features exactly characterize all them in general or what they have in common. Then we are considering semantic analysis. Moreover, we have to pay special attention to the analysis of the motion semantics, especially to the representation of movement of spatial entities and the associated reasoning. All these spatial objects that change their position or shape over time are called moving objects [14].

The analysis of motion semantics has become in an impor-tant issue for the correct representation and subsequent rea-soning about moving objects in the world (trajectory analysis, movement patterns recognition, movement prediction). When we represent planes or vehicles for tracking purpose, we can talk about moving points. For meteorologists hurricanes (that move or change their extension) can be represented by moving

regions. Finally, when we represent or track something like a moving fleet of ships or convoys, which can increase or diminish their extension, we have then a moving line. Most of these application fields are extremely important and some of them need accurate information in order to avoid unwanted disasters, for example aircraft tracking.

Some important proposals are: [16], [17], [10], [3], [13], [12], [19], [1], which we classify as foundational or visionary papers. Many of these papers deal with temporal and spatial semantics separately, even when describing the present prob-lem. On the other hand, there is another point of view for the representation and reasoning of spatiotemporal objects seman-tics: using logical and mathematical analysis (e. g. description logics, fuzzy, math algorithms, qualitative reasoning).

In this work we present an adaptation of a GIS application that have the capability of reasoning and inference using an ontological model. The proposed model helps on enhancing the obtained results in a heavily critical field related to Hurricane tracking and analysis.

This document is structured as follows: After the introduc-tion, section II describes the state of the art and related works. Then, section III explains the data model characteristics and importance. In section IV, the current proposal is presented, describing the reasoning process ans the system main com-ponents and, finally, section V shows conclusions and some aspects for been considered in the future.

II. STATE OF THE ART

One common direction followed by researchers and scien-tists today is to use Qualitative Reasoning (QR) and Qualitative Spatial Reasoning (QSR). In this last group is the Region Connection Calculus (RCC). The original idea of this work is found in [11], but we will focus on [4]. In RCC Cohn et al. take regions as primary spatial elements. RCC contains axioms that allow us to deal with spatial and physical objects, analyzing their relationships (e.g. containment relationships) and shape. Concerning moving objects, there is an approach called reasoning about continuous change, which tries to explore the semantic inside the relationship between regions (e.g. inside, outside).

This is an interesting way of tracking the approximate movement of regions using a set of snapshots of the relation-ships between them and analyzing the continuity. The main problem here is the lack of temporal semantics consideration, but for the mentioned Interval Temporal Logic, given that the authors consider that it closely mirrors the region-based spatial reasoning presented in RCC. Moreover, the complete
abandonment of the point-based reasoning leaves out one critical element for the analysis of moving objects: moving points. On the other hand RCC provides many

useful axioms, mainly associated with relationships between regions, which should help in the development of a future improved frame-work for the analysis of evolving entities trajectory, location management and other geographical reasoning, all this given the region-like shaped representation of spatial entities on a map (e.g. forest, buildings, cities).

The Qualitative Constraint Calculi (QCC) is another kind of qualitative reasoning. It has been applied in some areas for qualitative spatial reasoning, representing binary relationships between spatial elements like constraints and analyzing consis-tency or satisfiability of a given fact, and also for discovering information implicit in constraints. One clear example of the usage of QCC for our purpose is [7], where Gantner introduces a useful tool based in the Qualitative Constraint Calculi.

Finally, the most interesting and recent paper concerning the area of representation and reasoning about trajectories and tracking objects is The Hybrid Model [18]. The main goal that Yan et al. planned to achieve was to have a generic model that allows enriching semantically the trajectories obtained from GPS or any other mobility feeds. For this purpose, they provide a computing platform that enable the semantic enrichment of mobility data. Three structural logic components (Data Model, Conceptual Model and Semantic Model) are in charge of analyzing GPS feed sequences associating them by time intervals, establishing correlations between some features and finally adding knowledge that could be, for example, socio-political or geographical information. The obtained semantic concepts of mobility data could be used for reasoning in an ontology-based approach. The concept of ontology became stronger in the 90's, when Gruber defined ontology as an "explicit specification of a conceptualization" [9].

From that moment, ontologies became an important ele-ment inside the Artificial Intelligence field and still are today. Ontologies were first proposed in order to provide a common vocabulary for communication between systems and systems with persons. Over the years, they have become useful in many other application scenarios, including some approaches in the spatial and temporal field. The surging of ontology reasoners and rule based approaches have enabled the possibility of creating new information based on inferences using existing one. This heavily helps on enhancing the further analysis.

In this vein, some works for been considered are: [8], where Grenon presents the basic elements of a framework for spatio-temporal knowledge representation and reasoning. This framework helps on facing endurants and perdurants diverse modes of existence. Separated modules consider the semantics of both classifications. Some given axioms consider mereological and mereotopological features, intra-ontological and trans-ontological relationships. The establishment of these relationships between ontologies is an interesting goal for this proposal, given that we too consider that the integration of existent ontologies is a viable solution to the existing problem; [6], where Frank establishes a discussion about spatio-temporal proposed ontologies and methods for constructing them and their limitation and problems, aiming to find better proposals, suggest improvements and find a consistent ontology for spatio-temporal GIS. The proposed ontology deeply analyzes metric and topological properties, projections, snapshots and other topological and mereological elements and [15], where a clear idea of an ontological representation for moving objects is given. The main purpose of Tryfona and Pfoser is to allow interoperability between moving objects applications and, even so, important goals are achieved. They also propose architectural issues for mobile ontology-based applications. This can be considered a starting point for further development. Based on this approach, some improvement is needed, mainly in those aspects considering regions mobility and moving lines cases.

Related to GIS area is the rising of the Ontology Driven Geographical Information Systems (ODGIS), as stated in [5], with a tremendous importance. But most of existing systems only consider spatial and geographical semantics (with all as-pects it involves), nothing included from the needed movement semantics, totally focused in the spatio-temporal domain. The solution could be to think in an enhanced kind of ODGIS, starting from new spatio-temporal ontologies for the moving objects domain, not only the existing spatial or geographical ones.On the other hand, no intelligent-driven approach is assumed, which have a lack on accuracy and applicability.

III. THE LAYING DATA MODEL

As data model we propose an ontology that:

i. Captures the conceptualization of moving objects domain.
ii. Establishes a way for representing moving objects for subsequent recognition and classification (on data processing if needed) using the domain concept hier-archies and SWRL inference rules.

iii. Warrantee ontology interoperability by importing and mapping concepts and elements from existing formal ontologies.

For enclosing the whole domain of moving objects def-initions (also associated to representation and reasoning) it was important to entail a variety of widely spread concepts associated to moving objects, trajectories, movement and some other issues related to motion semantics. Some of the most important considered concepts are:

Spatial Entity: Represents any spatial or geographical entity and concept.

Temporal Entity: Represents temporal aspects, like instant and interval basically.

Moving objects: Encloses moving points, regions and lines.

Trajectory: Represents both spatial and semantic trajecto-ries.

Location: Represents geographical position or place.

On the other hand, there are some other elements as-sociated to the movement semantics, which in a certain way have more or less influence over the moving objects behavior. For example, one common classification of moving objects present in the bibliography is considering the environment restriction, and it is considered in our ontology (MovingObjectInConstrainedEnvironment, MovingObjectInUnconstrainedEnvironment, MovingObjectIn-Network), which is taking into account some movement and environment characteristics more than moving object itself.

Other important element we consider here (and is not considered on most previous ontologies) is the usage of SWRL rules for Allen´s Interval and RCC8 calculus for spatial and temporal relations between spatial entities respectively. We follow some elements given in [2], and fully apply the mentioned composition tables for both, Allen´s algebra and RCC.

In OntoMov, we follow this approach but enriching it semantically considering some other elements: continuity net-works, for reasoning over continue movement, and considering moving objects domain beyond a perdurantist approach, mix-ing both evolving and non-dynamic entities. Some rules for Allen's algebra are:

meets(?a, ?b), equal(?a, ?c) ! meets(?c, ?b)

finishes(?c, ?a), meets(?a, ?b) ! meets(?c, ?b)

during(?a, ?b), meets(?b, ?c) ! before(?a, ?c)

For spatial reasoning we introduce some RCC rules:

ExternallyConnected(?b, ?c), NonTangentialProperPart(?a, ?b) ! DisConnected(?a, ?c)

DisConnected(?b, ?c), TangentialProperPart(?a, ?b) ! DisConnected(?a, ?c)

For analysing moving regions we choose some RCC8 Continuity Networks based rules:

MovingRegion(?a), NonTangentialProperPart(?a, ?b), Par-tiallyOverlaps(?a, ?b) ! TangentialProperPart(?a, ?b)

MovingRegion(?a), ExternallyConnected(?a, ?b), Tangen-tialProperPart(?a, ?b) ! PartiallyOverlaps(?a, ?b)

There are growing advantages in the usage of rule-based reasoning over this ontology, which greatly enhance the on-tology expressivity and reasoning capacity.

Some emerging approaches, like the Rule Interchange Format (RIF[1]), will help on enhancing ontology reasoning and, much better, enhance spatio-temporal reasoning using rules.

IV. REASONING OVER GIS APPLICATIONS

We propose, as analysis tool, a GIS with enhanced ca-pabilities for analyzing the aforementioned data model. The GIS have a plugin-based communication with the ontological model and also with the reasoning engine, based in the Pellet reasoner.

Then, we have four main components:

- The ontological model
- The semantic extractor
- The reasoning engine
- The visualizer (over maps)

The ontological model, as explained before, is the support for semantic and spatio-temporal representation. The spatio-temporal information is gathered from the database and stored in the data model by the semantic extractor, which transform all spatial and spatio-temporal entities and relations to semantic ones, creating concepts and relation between concepts and also adding individuals to the laying ontology.

When some kind of processing of the information is needed for analysis, the reasoning engine process all the information in the ontology, generating new knowledge by ontology inference. This new knowledge can be used as a basis for a better analysis, ensuring more precise results. In the inference process the SWRL stated rules play an important role, given are the heart of the reasoning engine and the main source for generating new knowledge and information and, more important even, for using the new knowledge in the final analysis result.

We consider in this work some kind of analysis, based in the gathered information:

---

1   http://www.w3.org/TR/2010/WD-rif-overview-20100511/

- Hurricane classification
- Region classification
- Hurricane semantic trajectory analysis
- Hurricane impact evaluation

Hurricane classification helps in determining, by having the number of regions affected and other meteorological data like wind speed and rain, the hurricane category beyond common scale, more related to the level of destruction and caused damage. In the analysis, the inferred information is also used. Fig 1 illustrates this functionality over a GIS application. Using a color-based scale, we can assign a classification to a given hurricane (from low to high destructive hurricane).

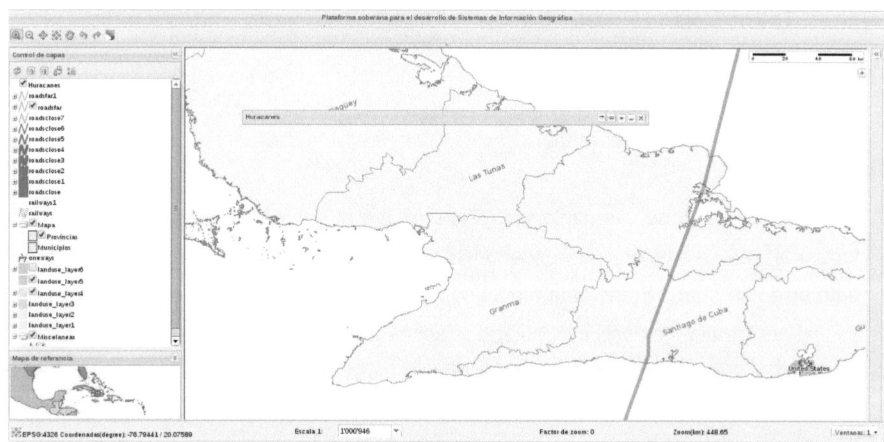

Fig. 1. Hurricane classification

Other analysis commonly needed is the classification of regions based in some aspects like: the number of hurricanes that affected them, the real impact of hurricanes over them, meaning destruction, number of deaths and more. Most of this kind of analysis and classification are human reasoning-like, which makes the reasoning engines evolve for obtaining as closest results as possible to human understanding. Here is where the actual proposal introduces a significant goal. Figure 2 shows the classification process for a given region.

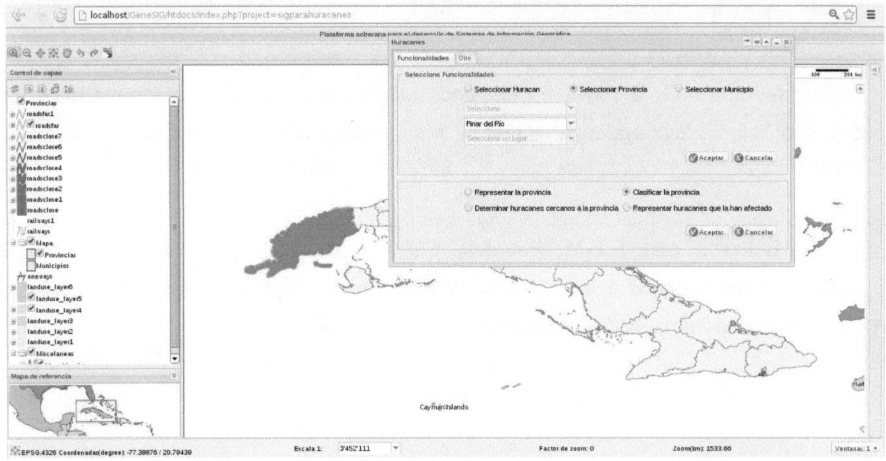

Fig. 2. Region classification

V. CONCLUSIONS

Even the pressing necessity of more accurate systems for predicting and analysis hurricane behavior, existing systems fall short when dealing with challenges like reasoning as humans can do. Geographical Information Systems, as the key for spatial and spatio-temporal analysis of those meteorological events, have to include desired capabilities. With our proposal we explode the possibility of generate new spatial and spatio-temporal information for using it in further analysis, which helps in better results and also understanding of the phe-nomena. Human-like reasoning is an important element to be considered in many research and application fields, moreover in GIS domain for spatial and spatio-temporal analysis. Some other elements like fuzzy logics can also be included in future proposals.

REFERENCES

[1] Alessandro Artale, Christine Parent, and Stefano Spaccapietra, Evolving objects in temporal information systems, Annals of Mathematics and Artificial Intelligence 50 (2007), 5–38.

[2] S. Batsakis and E.G.M. Petrakis, SOWL: spatio-temporal representa-tion, reasoning and querying over the semantic web, Proceedings of the 6th International Conference on Semantic Systems, ACM, 2010, pp. 1–9.

[3] Thomas Bittner, An ontology for spatio-temporal databases, 2001.

[4] A.G. Cohn, B. Bennett, J. Gooday, and N.M. Gotts, Qualitative spatial representation and reasoning with the region connection calculus, GeoInformatica 1 (1997), no. 3, 275–316.

[5] Frederico T. Fonseca, Max J. Egenhofer, Peggy Agouris, and Gilberto Camara,ˆ Using ontologies for integrated geographic information sys-tems, Transactions in GIS 6 (2002), no. 3, 231–257.

[6] A. Frank, A linguistically justified proposal for a spatio-temporal on-tology, Proceedings of the COSIT03 international conference. Position paper in COSIT-03 workshop, 2003.

[7] Z. Gantner, M. Westphal, and S. Wolfl, GQR–A fast reasoner for binary qualitative constraint calculi, Proc. of AAAI, vol. 8, 2008.

[8] P. Grenon, The Formal Ontology of Spatio-Temporal Reality and its Formalization, AAAI Spring Symposium on the Foundations and Applications of Spatio-Temporal Reasoning, 2003.

[9] T.R. Gruber et al., A translation approach to portable ontology speci-fications, Knowledge acquisition 5 (1993), 199–220.

[10] Dieter Pfoser, Indexing the trajectories of moving objects, IEEE Data Engineering Bulletin 25 (2002), 3–9.

[11] D.A. Randell, Z. Cui, and A.G. Cohn, A spatial logic based on regions and connection, KR 92 (1992), 165–176.

[12] John F. Roddick and Brian G. Lees, Paradigms for spatial and spatio-temporal data mining, (2001).

[13] Stefano Spaccapietra, Christine Parent, Nadine Cullot, and Christelle Vangenot, On Using Conceptual Modeling for Ontologies, Proceedings of the International Workshop on Intelligent Networked and Mobile Systems, Track 1: Ontologies for Networked Systems (ONS), in con-junction with the 5th International Conference on Web Information Systems Engineering WISE 2004, LNCS, Springer, 2004, pp. 22–33.

[14] E. Tøssebro and R. Guting, Creating representations for continuously moving regions from observations, Advances in Spatial and Temporal Databases (2001), 321–344.

[15] N. Tryfona and D. Pfoser, Designing Ontologies for Moving Objects Applications, Proc. Og Int. Workshop on Complex Reasoning on Geographic Data, Citeseer, 2001.

[16] Nico Van de Weghe and Philippe De Maeyer, Conceptual neigh-bourhood diagrams for representing moving objects, Perspectives in Conceptual Modeling (Jacky Akoka, Stephen Liddle, Il-Yeol Song, Michela Bertolotto, Isabelle Comyn-Wattiau, Samira Si-Said Cherfi, Willem-Jan Heuvel, Bernhard Thalheim, Manuel Kolp, Paolo Bresciani, Juan Trujillo, Christian Kop, and Heinrich Mayr, eds.), Lecture Notes in Computer Science, vol. 3770, Springer Berlin / Heidelberg, 2005, 10.1007/11568346 25, pp. 228–238.

[17] Ouri Wolfson, Moving objects information management: The database challenge, Next Generation Information Technologies and Systems (Alon Halevy and Avigdor Gal, eds.), Lecture Notes in Computer Science, vol. 2382, Springer Berlin /

Heidelberg, 2002, 10.1007/3-540-45431-4 7, pp. 15–26.

[18] Z. Yan, C. Parent, S. Spaccapietra, and D. Chakraborty, A Hybrid Model and Computing Platform for Spatio-semantic Trajectories, The Semantic Web: Research and Applications (2010), 60–75.

[19] Zhixian Yan and Stefano Spaccapietra, Towards Semantic Trajectory Data Analysis: A Conceptual and Computational Approach, VLDB2009 (PhD Workshop).

YOUR KNOWLEDGE HAS VALUE

- We will publish your bachelor's and master's thesis, essays and papers

- Your own eBook and book -
  sold worldwide in all relevant shops

- Earn money with each sale

Upload your text at www.GRIN.com
and publish for free